MW00444337

SLUGS AND SNAILS

CREEPY CRAWLERS

Lynn Stone

The Rourke Book Co., Inc.
Vero Beach, Florida 32964

PHOTO CREDITS
Title page © Breck Kent; page 15 © James A. Robinson; all other photos © Lynn M. Stone

Library of Congress Cataloging-in-Publication Data

Stone, Lynn M.
Snails and slugs / by Lynn M. Stone.
p. cm. — (Creepy crawlers)
Includes index.
Summary: Describes the physical characteristics, behaviour, and life cycle of snails and slugs.
ISBN 1-55916-159-0
1. Snails—Juvenile literature. 2. Slugs (Mollusks)—Juvenile literature. [1. Snails 2. Slugs (Mollusks)]
I. Title II. Series: Stone, Lynn M. Creepy crawlers
QL 430.4.S85 1995
594'.3—dc20 95–18458
CIP
AC

Printed in the USA

TABLE OF CONTENTS

SNAILS AND SLUGS

Those hard, fancy sea shells you find on seashores once belonged to soft, not-so-fancy snails.

Snails and their cousins, the slugs, are simple, slimy animals. They are legless, so most of them crawl. Some **species** (SPEE sheez), or kinds, swim or float.

Snails have the same basic bodies as slugs, but snails wear shells. Snails make their shells from a body of liquid that hardens. Long after a snail dies, its shell survives.

Slugs are close cousins of snails, but slugs don't have a shell

MOLLUSCS

Snails and slugs belong to a group of animals called **molluscs** (MAH luhsks). Molluscs are **invertebrates** (in VERT uh brayts), animals without bones.

Scientists have identified over 110,000 species of molluscs. Ninety thousand are snails and slugs! Some of the other molluscs are clams, oysters, squids, and octopuses. The giant squids are the largest of all invertebrate animals.

The molluscs are the largest group of **marine** (muh REEN), or saltwater, animals without bones.

Slugs are molluscs without shells

WHAT SNAILS AND SLUGS LOOK LIKE

Snails and slugs have fleshy bodies with a pair of **antennas** (an TEN uhz) on their head. The sensitive antennas can be raised like rabbit ears. They help an animal know what is going on around it.

Some of the snails have a muscular "foot." The foot helps the animal crawl and attack other animals.

Slugs are generally brownish or greenish. The banana slug, however, is yellow. Most snails have dull shells, but a few are spotted or brightly striped.

This Bahama tree snail shows a snail's antennas (upper left) and eye stalks (upper right)

HOW SNAILS AND SLUGS LIVE

Whether they crawl, swim, or float, snails and slugs take their time. Moving "at a snail's pace" means moving slowly indeed.

Although slow, snails and slugs are not helpless. Snails are protected somewhat by their shells. Some species have a tough, leathery **operculum** (o PERK u lum) at the base of the foot. When the foot is withdrawn, the operculum, like a door, blocks the shell entry.

An olive snail burrows into sand

Collectors prize some shells for their bright colors and fancy design

Eggs of the freshwater apple snail are clustered on a leaf of water lettuce

WHERE SNAILS AND SLUGS LIVE

Snails and slugs live on land, in the sea, and in bodies of fresh water. Some of the most familiar little slugs live in gardens and woodlands. They leave a trail of **mucus** (MU kuhss) that oozes from their skin.

Snails live in unlikely places. One species lives in the Sahara Desert. Another lives in the trees of Everglades National Park in Florida.

Brightly colored Everglades tree snails live in south Florida

WHAT SNAILS AND SLUGS EAT

What a snail or slug eats depends upon the kind of snail or slug. Land snails and slugs, for example, graze on plants. Certain marine snails are meat-eaters. With the strong foot, these snails open clams and oysters and feast on the meat inside.

Several common American marine snails, such as the tulip snail, attack and eat other snails.

Powerful foot and sharp operculum help make the tulip snail a predator of other marine snails

ENEMIES

Crawling about without shell armor is not always safe. Snails have enemies, large and small.

The hawklike Everglades kite, for example, is a snail-eating expert. The kite's thin, sharply hooked beak is perfect for plucking apple snails out of their brown shells. Another bird, the limpkin, uses its long, narrow beak for the same purpose.

Crabs eat some marine snails, and mites and small insects attack tree snails and slugs. Water pollution also kills snails.

Only this apple snail's shell was left after the animal inside was plucked by a limpkin

YOUNG SNAILS AND SLUGS

Most snails and slugs lay eggs. A tiny snail or slug hatches from each egg.

Some of these molluscs produce a long, leathery string of egg "pockets." Their youngsters hatch from these pockets. The best known of these egg strings in the United States is the lightning whelk snail's. Thousands of egg cases from this marine snail wash onto Atlantic Ocean beaches.

A lightning whelk's egg case lies on an Atlantic beach

SNAILS, SLUGS, AND PEOPLE

Certain snails have been popular as people food for many years. Some restaurants serve snails under a French name—**escargot** (ehs kar GO).

In the Florida Keys and Caribbean Sea region, conch snails are served—without the shells, of course—in chowders, salads, and cakes.

Some snails are prized for their colorful and delicate shells. Unfortunately, shell collecting has reduced the number of many snails, such as the Everglades tree snail.

Glossary

antennas (an TEN uhz) — on the heads of snails and slugs, stalklike structures used to sense smell, movement, and location

escargot (ehs kar GO) — snails prepared for human food

invertebrates (in VERT uh brayts) — the simple, boneless animals, such as worms, snails, starfish, and slugs

marine (muh REEN) — of or relating to the sea

molluscs (MAH luhsks) — simple, boneless animals, including many with shells, such as oysters, clams, and snails

mucus (MU kuhss) — the sticky liquid produced by snails and slugs and certain other animals

operculum (o PERK u lum) — a hard covering, like a shoe, on the bottom of the foot of certain snails

species (SPEE sheez) — within a group of closely related animals, one certain kind, such as a *banana* slug

INDEX